HASAN IBN AL HAYTHAM

HASAN IBN AL HAYTHAM

(965-1040)

THE FIRST SCIENTIST

M. BASHEER AHMED, M.D.

Abu Ali Hasan Ibn al-Haytham – Alhazen
(965-1040)
The First Scientist

M. Basheer Ahmed, M.D.

ISBN (Print): 978-1-966561-39-2
ISBN (e-Book): 978-1-966561-40-8

Published by
Mary Ethel Eckard
Frisco, TX

This is part of the series of books on Muslim scientists of the medieval era.

CONTENTS

FOREWORD

BY BRADLEY STEFFENS

THE HUMBLE MATHEMATICIAN WHO CHANGED THE WORLD

Every few years, someone will publish a list of the "100 most influential books of all time." Certain religious books—the Qu'ran, the Bible, and The I Ching—usually are included, because they have shaped the daily lives of billions of people throughout history. Poems, plays, and works of fiction have influenced human history as well. Lists of important books usually include The Works of William Shakespeare, *The Iliad* and *The Odyssey* by Homer, and *The Divine Comedy* by Dante Alighieri. Scientific works are often included, too. A 2015 list published by the World Economic Forum put Charles Darwin's *On the Origins of Species* as the most influential book in history. So far, however, none of these list makers have included what is arguably the most important scientific work of all time, *The Book of Optics* by the eleventh century Islamic mathematician Alhasan Ibn al-Haytham.

Why is *The Book of Optics* so important? It is not because of the discoveries that Ibn al-Haytham describes in it, although they are important. Ibn al-Haytham was the first person to realize how vision occurs and how light travels through the air and other media. These discoveries paved the way for the entire science of optics. But what is more important is the way that Ibn al-Haytham proved the truth behind his discoveries.

Before Ibn al-Haytham wrote *The Book of Optics*, scientists and philosophers established the truth through logical reasoning. While he was working on *The Book of Optics*, Ibn al-Haytham found that earlier scholars had been wrong about many things regarding light and vision. He realized that logic alone is not enough to establish the truth about nature. Ideas about how nature works must be tested in the physical world to see if they are true. He developed concrete ways to test his ideas. He called these tests "true demonstrations." Today, we call them experiments.

Ibn al-Haytham realized that his ideas about true demonstrations applied not only to the works of others, but to his own work as well. "I formerly composed a treatise on light and vision in which I employed persuasive methods of reasoning," he wrote, "but when true demonstrations relating to all objects of vision occurred to me, I started afresh. Whoever, therefore, comes upon the said treatise must know that it should be discarded." He declared that he would not include anything in *The Book of Optics* unless it had been proven with mathematics or a true demonstration. This is the methodology that all scientists have followed ever since. *The Book of Optics* was translated into Latin and spread through

Europe, inspiring new generations of scientists to not only reason, but to test their ideas through experiments. This is known as the scientific method.

Ibn al-Haytham gave not just scientists, but regular people, a new way to know the truth about nature. Today, in courts of law, juries often are not swayed by arguments alone. They want physical proof. When an advertiser says a product works better or tastes better than another, consumers want proof, and products are tested. That, too, is a form of science. And it all started with Alhasan Ibn al-Haytham, the first true scientist.

In this concise biography, M. Basheer Ahmed, M.D. masterfully describes the colorful events in Ibn al-Haytham's life and how they shaped his journey of discovery. He also outlines the important discoveries Ibn al-Haytham made and why he is considered the first modern scientist.

Bradley Steffens is the author of seventy-five nonfiction books for young adults, including *Ibn al-Haytham: First Scientist* (2007) and *First Scientist, Ibn al-Haytham* (2020). He also has written two young adult novels based on the life of Ibn al-Haytham: *The Prisoner of Al-Hakim* (2017) and *The Empty Quarter* (2025).

PREFACE

By Dr. M. Basheer Ahmed, M.D.

Light, Science, and a Millennium-Old Legacy

Western historians often highlight the achievements of Greek and Roman scholars up to around 300 AD, then leap forward to the 16th-century Renaissance—labeling the 1,300 years in between as "The Dark Ages of Europe." What is rarely acknowledged is that this period was anything but dark for the Muslim world. It was an age of extraordinary brilliance, during which Muslim scientists and scholars illuminated human civilization with groundbreaking contributions in mathematics, physics, chemistry, medicine, astronomy, social sciences, and philosophy.

Inspired by the Qur'anic command to seek knowledge, Muslims established great centers of learning such as Bayt al-Hikma in Baghdad and universities stretching from Central Asia to Cordoba. Giants like Ibn al-Haytham, Ibn Sina, Al-Razi, Al-Zahrawi, Ibn Rushd, Al-Farabi, Ibn Khaldun, and Ibn al-Nafis reached unparalleled heights in every field of inquiry. Their discoveries and writings later became the intellectual foundation upon which the European Renaissance was built.

Yet, over time, certain theologians and political leaders discouraged the study of science and philosophy. By the 15th century, no major university in the Islamic world continued to teach these disciplines. Today, many Muslims—especially students—know little about the remarkable scholars of the medieval Islamic era or the profound impact they had on world civilization.

It is therefore our responsibility to revive this knowledge and share it widely. The legacy left by these scholars is not merely a record of past glory; it is a treasure meant to inspire future generations. Their achievements should ignite pride, curiosity, and ambition in young Muslims, motivating them to pursue knowledge, innovate, and contribute to humanity.

This series of biographies is a humble effort to bring the luminous contributions of medieval Muslim scholars back into the public consciousness.

A Message to Young Muslims

The Muslim world once led humanity in mathematics, astronomy, medicine, and philosophy. That leadership can be revived—but only through education, critical thinking, and a renewed commitment to excellence.

Your ancestors shaped the world. Your heritage is one of science, innovation, and leadership. Your future can be just as bright, but it depends on the choices you make today.

Do not allow your history to be forgotten or distorted. Study it. Learn from it. Build upon it. The world urgently needs scientists, engineers, thinkers, and problem-solvers—young minds who understand that learning is empowerment and that knowledge is a path to dignity.

The legacy of Ibn al-Haytham is not a relic of the past; it is a call to action. His life teaches us that contributions can endure for centuries, but identities can fade if they are not protected and renewed. His example is a roadmap for today's youth: rise, study, question, innovate, and contribute with the same brilliance and dedication that once made Muslim civilization a beacon of science and learning.

The next chapter of this legacy belongs to you.

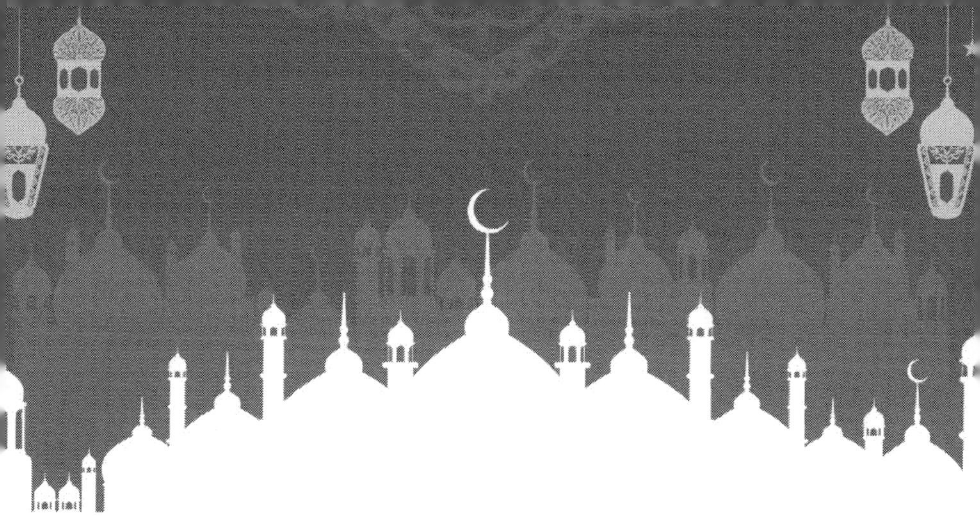

THE IMPORTANCE OF LIGHT-BASED SCIENCE AND TECHNOLOGY

In 2015, UNESCO declared the International Year of Light to emphasize the importance of light-based science and technology in modern life. Light plays a crucial role in energy production, communication, medicine, agriculture, and education. From lasers used in surgery to fiber-optic cables that carry the internet, light shapes our world in countless ways.

That same year also marked the 1,000th anniversary of important discoveries made by Ibn al-Haytham, a Muslim scientist whose work laid the foundation for the modern understanding of light and vision. Hundreds of years before scientists like Galileo and Newton, Ibn al-Haytham wrote a monumental scientific work called *Kitab al Manazir* (*Book of Optics*), which transformed how people studied nature and conducted scientific research. His work changed science forever

Ibn al-Haytham is remembered not only for his discoveries but also for how he made them. His insistence on experimentation and evidence earned him the title "the first true scientist."

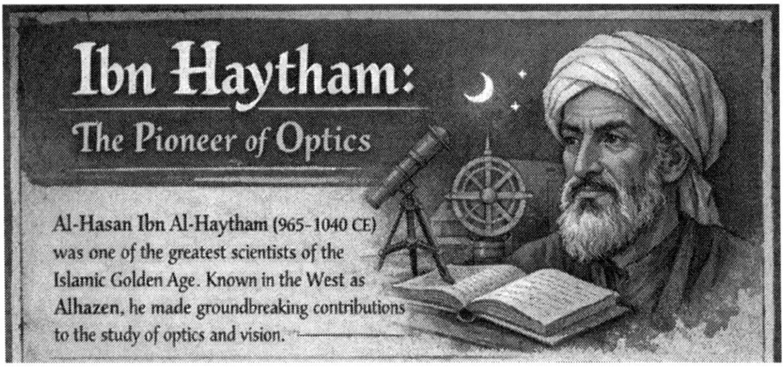

WHY IS IBN AL-HAYTHAM CALLED THE FIRST SCIENTIST?

Before Ibn al-Haytham's time, many scholars accepted ideas simply because famous thinkers from the past had proposed them. Questioning those ideas was rare and often discouraged. Ibn al-Haytham believed this approach was wrong. He believed that to find the truth, one must test ideas through physical evidence. He famously said: "If learning the truth is the scientist's goal, then he must make himself the enemy of all that he reads." This approach—forming a hypothesis, testing it, and accepting the results whether they confirm or contradict expectations—is now known as the scientific method.

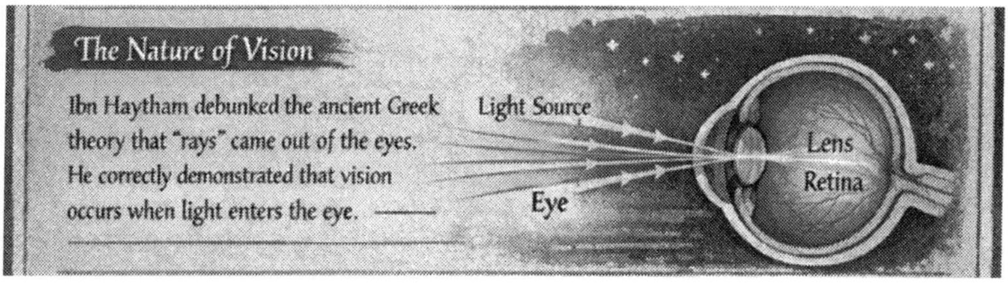

Ibn al-Haytham famously wrote that a seeker of truth must treat existing knowledge with skepticism and test it thoroughly. Because he consistently applied experimentation to scientific questions, historians of science regard him as one of the earliest practitioners of modern scientific thinking. Because he refused to accept "facts" without proof, historians recognize him as the first true scientist.

EARLY LIFE AND THE GOLDEN AGE OF LEARNING

Ibn al-Haytham was born in 965 CE in Basra, Iraq, during the Golden Age of Islamic civilization. This period was marked by extraordinary achievements in science, medicine, mathematics, philosophy, and engineering.

Major cities such as Baghdad, Cairo, Damascus, and Córdoba were centers of learning. Libraries held thousands of books, scholars debated ideas openly, and students studied subjects ranging from astronomy to medicine. Knowledge from Greek, Roman, Indian, and Persian civilizations was translated into Arabic, preserved, and expanded upon.

Growing up in this environment, Ibn al-Haytham developed a strong interest in learning. Although he initially worked as a judge, he soon grew dissatisfied with political and religious conflicts. He decided instead to devote his life to scientific inquiry where he could pursue truth without compromise.

A Scholar of Many Fields

In Europe, Ibn al-Haytham was known as Alhazen. He was a polymath, meaning he excelled in many areas of knowledge. His studies included: physics, optics, mathematics, astronomy and engineering.

Despite his wide interests, his most important contributions were in the field of optics, the study of light and vision. In the West, he is often known by the Latinized name Alhazen.

The Incident That Made Ibn al-Haytham a True Scientist

There was a problem of flooding in the Nile River in Egypt. Caliph al Hakim knew about Ibn al-Haytham of Basra, an engineer who wrote about building a dam on the Nile River to control flooding and preserve water for irrigation. The Caliph invited him to Cairo to regulate the flooding of the Nile and gave him a team of trained engineers to work with him.

Ibn al-Haytham soon realized that the project required more research and was not feasible due to technological limitations, especially because the width of the Nile doubled near Aswan.

The Egyptian Caliph, Al Hakim, was disappointed upon hearing this and Ibn al-Haytham feared the Caliph's harsh reputation. It is believed that he pretended to be insane to escape unjust punishment.

Ibn al-Haytham was placed under protective custody in his home until Al Hakim died in 1021. He remained under house arrest for about ten years. During this solitary period, he immersed himself in scientific research and experimentation. It was during this time that he made some of his most important optical discoveries.

While in his dark and isolated home, he noticed a tiny hole that allowed a stream of light to enter the room. He observed an astonishing phenomenon: an image of the outside world was projected onto the opposite wall. Fascinated, he continued experimenting with mirrors and lenses, trying to understand how the outside scene could be projected through a small hole onto his wall.

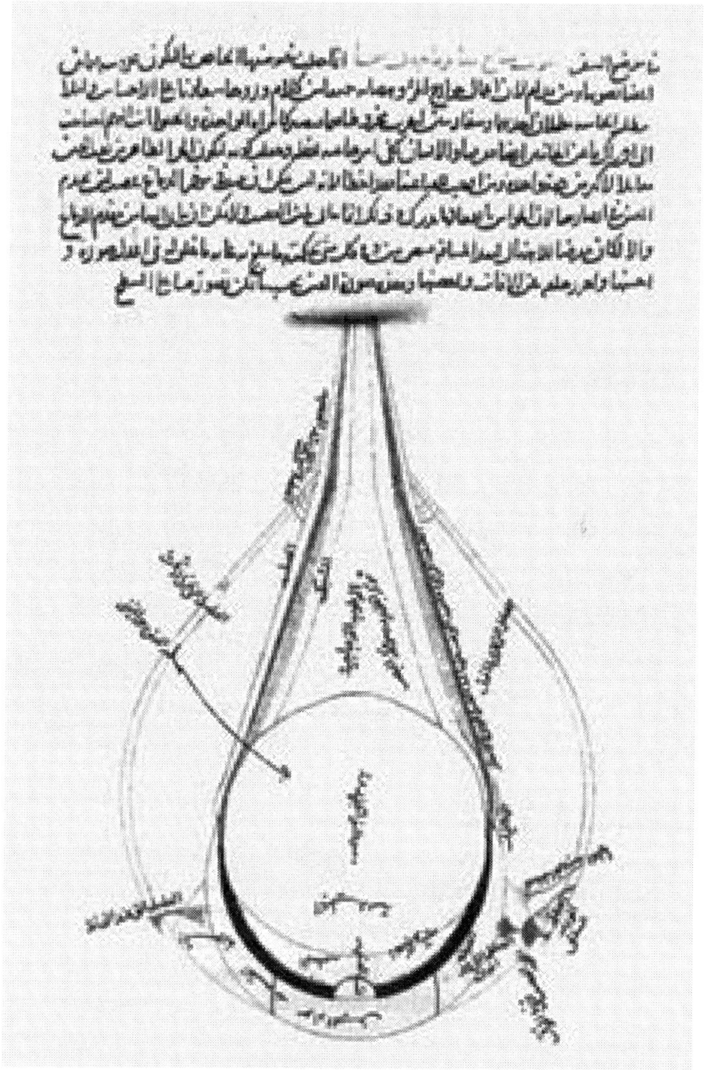

An anatomical diagram of the human eye by Ibn al-Haytham

After years of research, he concluded that the sun illuminated objects outside, and the light rays bounced off those objects in straight lines, directly to the eye. He set a new standard in experimental science and completed his great work on optics.

Changing the Understanding of Vision

For centuries, many scientists like Pythagoras, Euclid, and Ptolemy believed "Emission Theory" —that vision occurred because rays of light came out of the eyes, struck objects, and returned to the eyes. This idea was accepted without question.

Ibn al-Haytham challenged this belief. Through experiments, he showed that light travels from objects into the eye, not the other way around. He explained that objects reflect light which enters the eye and forms an image on the retina. The image is then sent through the optic nerve to the brain where it is processed.

This was a revolutionary idea. Ibn al-Haytham was also the first to explain that vision occurs in the brain, not in the eyes themselves. Today, this concept is fundamental to neuroscience and visual science.

Because of these discoveries, Ibn al-Haytham is widely called the "Father of Modern Optics."

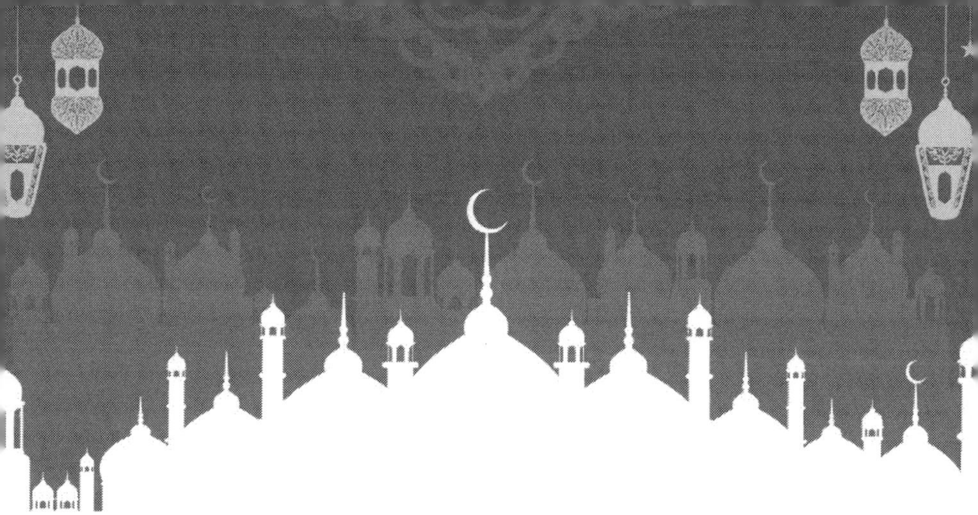

THE BOOK OF OPTICS: A SCIENTIFIC MASTERPIECE

Between 1011 and 1021, Ibn al-Haytham wrote *Kitab al-Manazir*, a seven-volume scientific work that became one of the most influential books in the history of science.

The Main Topics of the seven volumes are:

Volume I:	Nature of light, vision, and the structure of the eye
Volume II:	Visual perception
Volume III:	Conditions for clear vision and visual errors
Volumes IV–V:	Reflection and mirrors
Volume VI:	Errors in vision caused by reflection
Volume VII:	Refraction and lenses

Throughout the book, Ibn al-Haytham emphasized experimentation over theory. He tested ideas using controlled observations, measurements, and repeated trials.

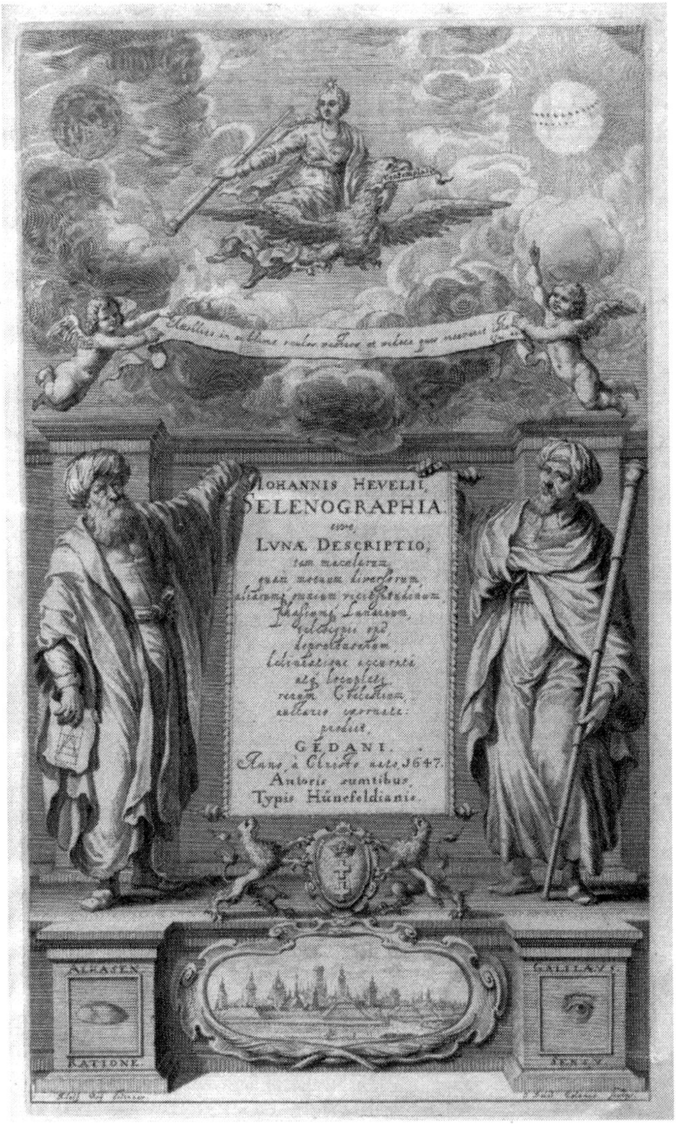

Ibn al-Haytham (left) and Galileo (right) depicted in an atlas of the moon

Light Travels in Straight Lines

One of Ibn al-Haytham's most important findings was that light travels in straight lines. He demonstrated this through experiments involving shadows, pinholes, and mirrors.

This insight helped explain why shadows have sharp edges and why images can be projected onto surfaces. It also became essential for later developments in optics, photography, and astronomy.

The Camera Obscura

While living in isolation in Egypt, Ibn al-Haytham noticed light entering a dark room through a small hole and projecting an inverted image of the outside world onto the wall.

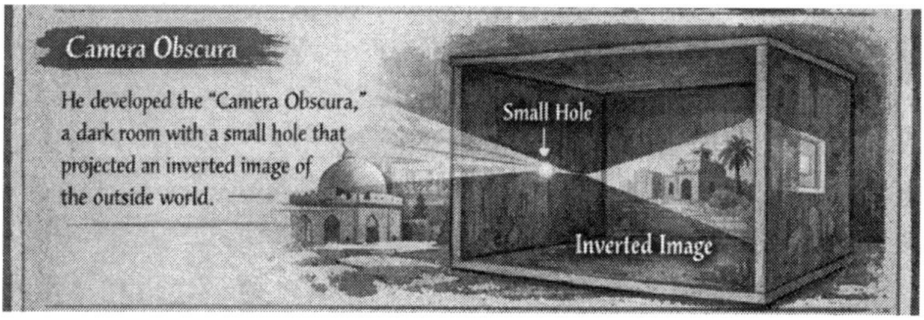

This observation led him to describe the camera obscura, meaning "dark chamber." This device demonstrated how light travels and forms images and later inspired the invention of the modern camera.

He even used the camera obscura to observe solar eclipses safely, protecting his eyes from damage.

Key Scientific Contributions

Ibn al-Haytham's lasting contributions include the following: introduction of experimental science; correct explanation of human vision; discovery of the principles of reflection and refraction; description of the camera obscura; study of optical illusions; identification of parts of the eye including the retina, lens, and cornea; division of light into primary light (from sources like the sun) and secondary light (reflected from objects).

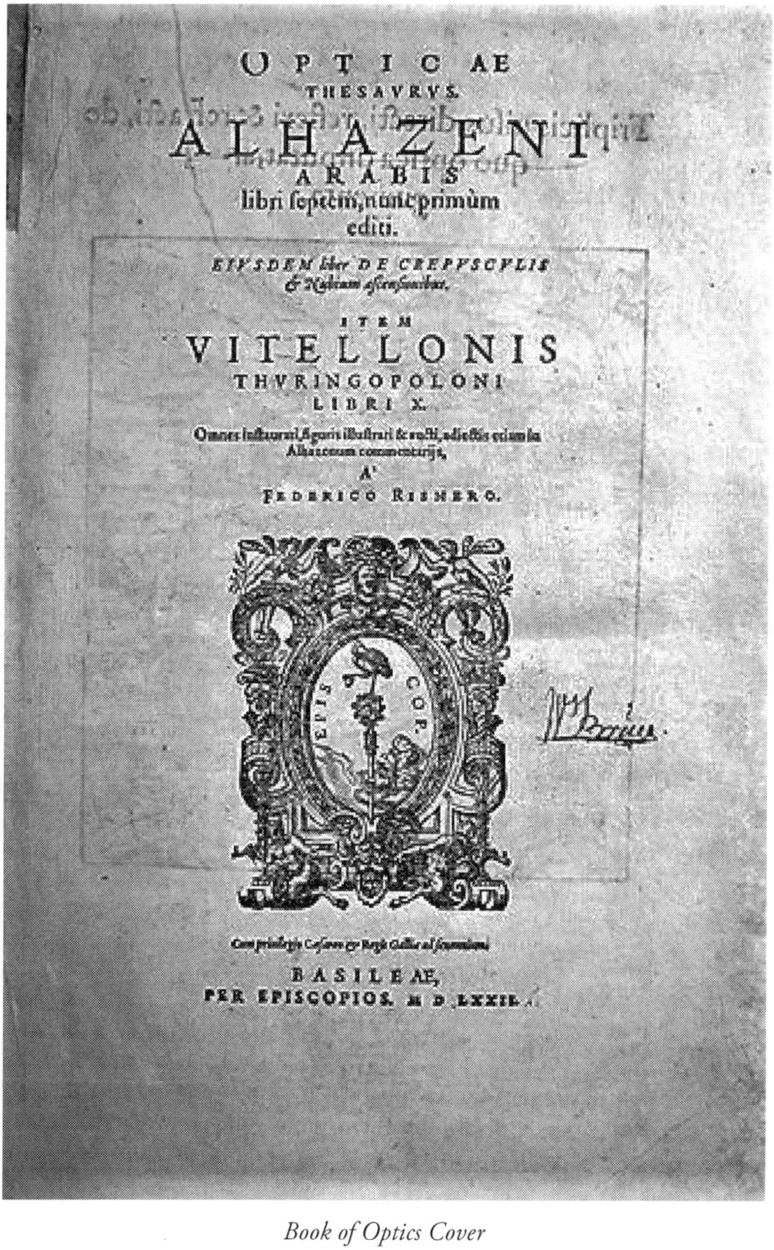

OPTICAE
THESAVRVS.
ALHAZENI
ARABIS
libri septem, nuncprimùm
editi.

EIVSDEM liber DE CREPVSCVLIS
& Nubium ascensionibus.

ITEM
VITELLONIS
THVRINGOPOLONI
LIBRI X.

Omnes instaurati, figuris illustrati & aucti, adiectis etiam in
Alhazenum commentarijs,
A'
FEDERICO RISNERO.

Cum privilegio Caesareo & Regis Galliae ad sexennium.

BASILEAE,
PER EPISCOPIOS. M D LXXII.

Book of Optics Cover

INFLUENCE ON EUROPE AND THE RENAISSANCE

In the 12th century, *The Book of Optics* was translated into Latin and became Europe's main textbook on optics for more than five centuries. It strongly influenced scientists such as Roger Bacon and Johannes Kepler.

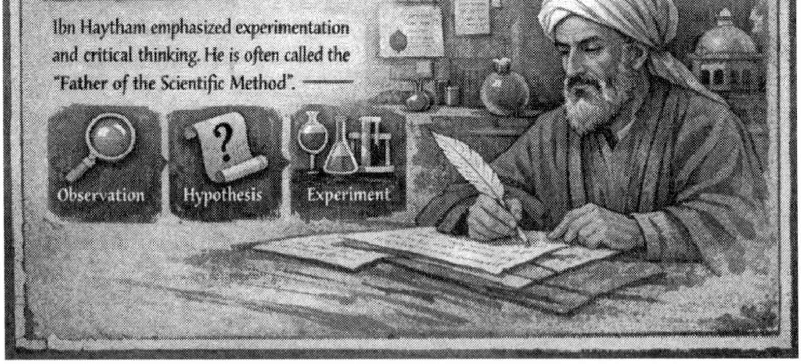

Through these translations, Muslim scholarship helped lay the groundwork for the European Renaissance, a period of renewed scientific and intellectual growth.

Ibn al-Haytham's writings were more influential in Latin than in Arabic. While European scientists appreciated and admired the scientific work and discoveries of great medieval Muslim scientists, Muslims themselves gradually abandoned the teaching of science and technology. By the 18th century, many Muslim students did not even recognize the names of great medieval Muslim scholars. In the late 19th and early 20th centuries, European and American historians reintroduced these scientists to the Muslim world, and much of Ibn al-Haytham's great work was translated from its Latin version of *The Book of Optics* (Kitab al Manazir).

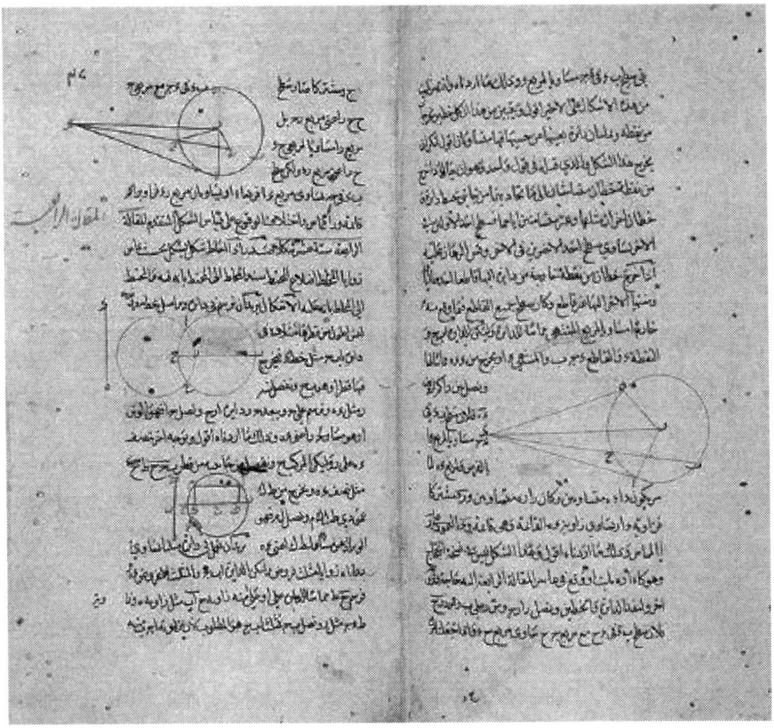

Page from the Book of Optics

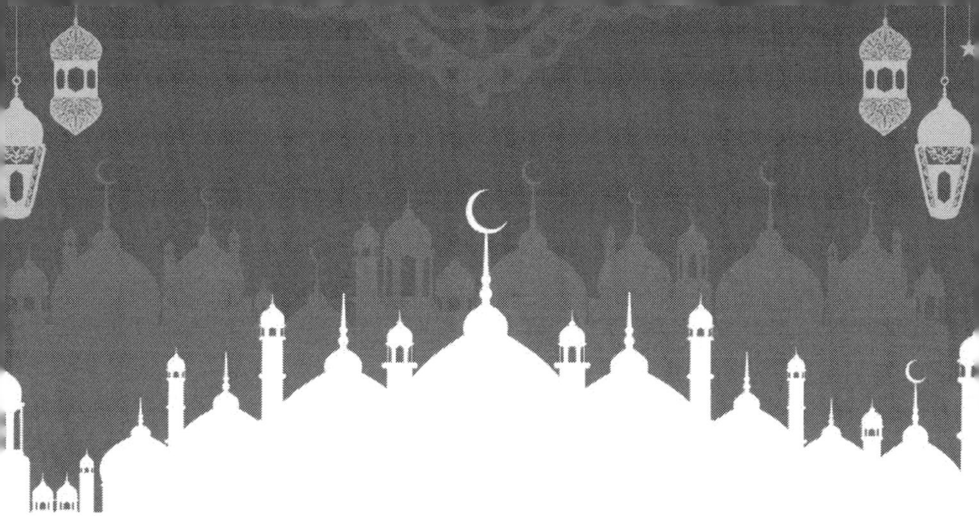

LEGACY AND RECOGNITION

Although Ibn al-Haytham wrote nearly 200 books, only about 65 have survived. His work remains deeply influential in modern science.

He is honored today through: The Alhazen crater on the Moon, Asteroid 59239 Alhazen.

Recognition as a Pioneer of Scientific Methodology.

Modern technologies such as cameras, lasers, medical imaging, and fiber optics are all built on principles he discovered over a thousand years ago.

Faith and the Pursuit of Knowledge

Ibn al-Haytham believed that Islam encourages the pursuit of knowledge and careful study of the natural world. His faith inspired him to seek truth through honesty, patience, and rigorous experimentation. His life demonstrates that faith and science can work together, each enriching the other.

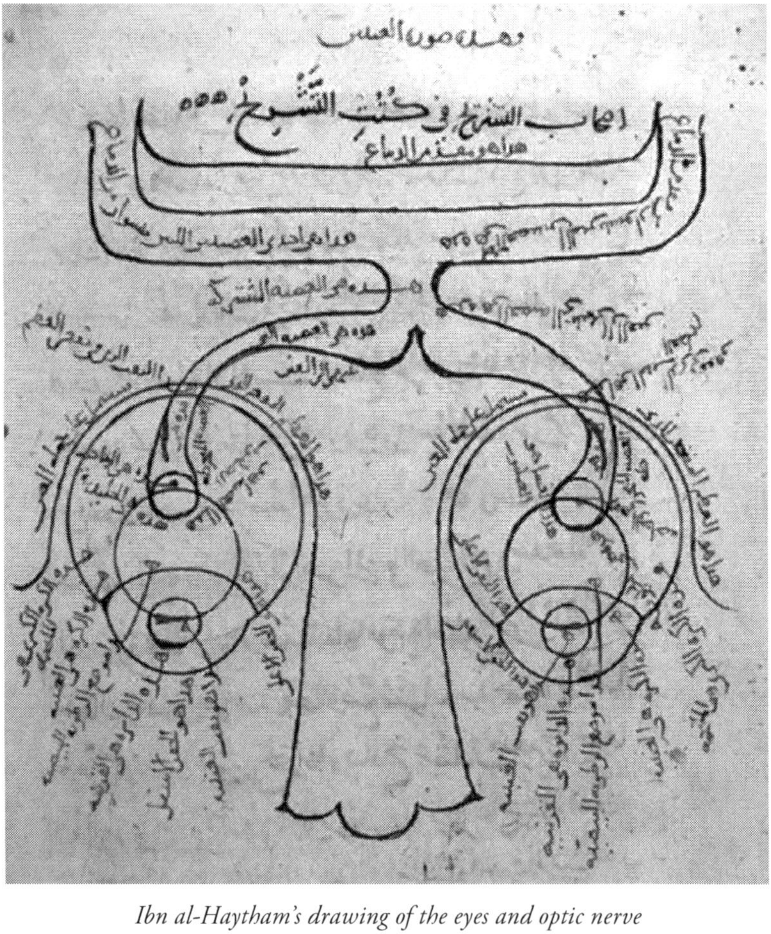

Ibn al-Haytham's drawing of the eyes and optic nerve

CONCLUSION

A Role Model for Future Scientists

Ibn al-Haytham's legacy teaches students that real progress comes from curiosity, courage, and critical thinking. He showed that questioning accepted ideas is not disrespectful, it is essential for discovery.

The history of science is more diverse than our textbooks often show. By looking back at pioneers like Ibn al-Haytham, we can see that great ideas don't belong to just one culture or time—they belong to anyone with the curiosity to chase them.

1. Don't Just Take Their Word for It

    Ibn al-Haytham lived a thousand years ago, but he was a total disruptor. At the time, people believed ancient Greek theories without question. He dared to say, "Show me the proof." He taught us that science isn't about memorizing facts; it's about having the courage to question "common sense" and test ideas for oneself.

2. Science and Belief Can Coexist

He showed that we don't have to choose between faith and logic. For him, studying the laws of light and nature was a way to better understand the world. This balance helps break down walls and shows that science is a home for everyone, regardless of their background or beliefs.

3. Resilience in the Dark

One of the most incredible parts of his story is that he did some of his best work while under "house arrest" in Cairo. Even when his world became small and dark, his mind stayed wide open. It's a reminder that perseverance is just as important as intelligence. One person can spark a revolution of ideas even in the toughest circumstances.

4. His Legacy is in Your Pocket

Every time we take a selfie, use a laser, or put on glasses, we're using Ibn al-Haytham's legacy. His book, *The Book of Optics*, changed how we understand vision. He proved that light enters our eyes rather than shooting out of them (which is what people used to think!).

5. A Call to Action: Evidence Over Rumors

Today, we are flooded with information and "fake news." Ibn al-Haytham's message is more relevant than ever:

Value evidence over dogma.
Be curious, not indifferent.
Ask "why" and "how."
Work together across borders.

Science is a global relay race. Ibn al-Haytham passed the torch a thousand years ago—now it's your turn to carry it toward the next big discovery.

The "Skeptic" Mindset: Ibn al-Haytham believed we should be "enemies" of the books we read—meaning we should stay skeptical and test every claim. In an age of social media and AI, how can we apply his "skepticism" to the information we see online today?

Innovation Under Pressure: Ibn al-Haytham conducted some of his most famous experiments while under house arrest. Do you think difficult circumstances make people more creative, or was he just a rare exception? Can you think of a modern example of "innovation through struggle"?

Science vs. Tradition: He dared to prove the ancient Greeks wrong, even though they were considered the ultimate authorities for centuries. Why is it so hard for society to let go of "established" ideas, and what "facts" today do you think might be proven wrong in the future?

The Ethics of Discovery: Ibn al-Haytham saw his scientific work as a way to grow closer to truth and his faith. Today, we often treat science and personal beliefs as two separate worlds. Do you think having a strong personal "why" (like faith or a desire to help humanity) makes someone a better scientist?

Field	**Discovery / Contribution**	**Impact on Modern Science**
Scientific Method	Introduction of **experimental testing** and controlled observation to prove theories.	Formed the basis of all modern scientific research and peer review.
Optics	**Intromission Theory**: Proved light reflects off objects and enters the eye.	Overturned centuries of incorrect Greek theories about how humans see.
Anatomy	Detailed description of the **eye's structure**, identifying the cornea, lens, and retina.	Laid the groundwork for ophthalmology and corrective lens technology.
Photography	Invention of the **Camera Obscura** (pinhole camera).	Directly led to the development of the modern photographic camera.
Neuroscience	Discovered that **visual perception** occurs in the brain rather than the eyes.	Established the early connection between optics and cognitive psychology.
Physics	Proved light travels in **straight lines** and studied refraction/reflection.	Essential for the development of telescopes, microscopes, and laser technology.

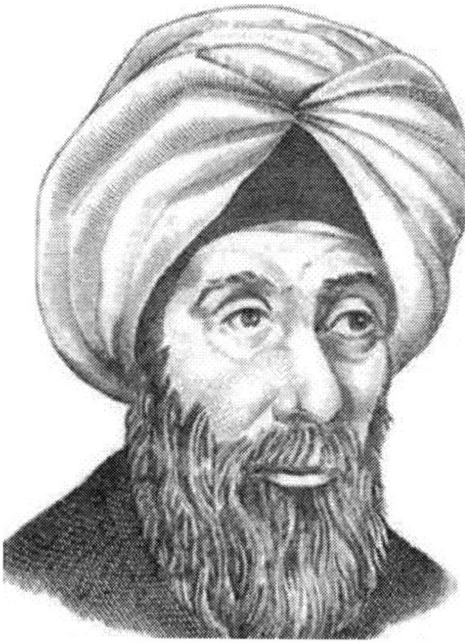

"The seeker after the truth is not one who studies the writings of the ancients ... but rather the one who suspects his faith in them and questions what he gathers from them."

—Ibn al-Haytham

This series of biographies is a humble effort to bring the luminous contributions of Medieval Muslim scholars back into the public consciousness.

BIBLIOGRAPHY

1. Bradley Steffens. *The Prisoner of al-Hakim*. Clifton, New Jersey. Blue Dome Press (2017)
2. Bradley Steffens. *First Scientist Ibn al-Haytham*. Clifton, New Jersey. Blue Dome Press (2021)
3. Turner, H. *Science in Medieval Islam*. Austin, Texas. University of Texas Press (1995)
4. Jim Al-Khalili, *Book of Optics*, by Ibn al-Ibn Al-Haytham.
5. https://www.nature.com/articles/518164a.pdf?origin=ppub
6. https://www.britannica.com/biography/Ibn-al-Ibn Al-Haytham
7. https://pubmed.ncbi.nlm.nih.gov/18822953/

QUESTIONS

Which theory of vision did Ibn al-Haytham disprove?

A) The idea that light reflects off objects into the eye.

B) The "Emission Theory" (that eyes send out invisible rays).

C) The idea that the brain processes images.

D) The theory that light travels in straight lines.

What did Ibn al-Haytham call the device that projected an upside-down image in a dark room?

A) The Telescope

B) The Retina

C) The Camera Obscura

D) The Pinhole Mirror

Part 2: True or False

True / False: Ibn al-Haytham believed that the eye itself, not the brain, was where vision was finally "seen."

True / False: He completed his most famous work, *The Book of Optics*, while under house arrest in Egypt.

True / False: Ibn al-Haytham's work was ignored in Europe and had no influence on the Renaissance.

Part 3: Critical Thinking & Discussion

The "Enemy" of the Text: Ibn al-Haytham said a scientist must make themselves the "enemy" of all they read. In our modern world of social media and "fake news," how can we apply his scientific method to the information we see online today?

Failure as a Catalyst: Ibn al-Haytham's engineering project on the Nile was a failure, and he ended up under house arrest. How did he turn this "failure" and period of isolation into one of the most productive times in scientific history?

The Anatomy of Vision: Using what you know about the Camera Obscura, why do you think the image projected on the back of our retina is actually upside-down? How does that relate to Ibn al-Haytham's discovery that the brain is responsible for "correcting" what we see?

Part 1: Multiple Choice (Answer Key)

Answer: C

Explanation: While he did study light and live in the Golden Age, his unique contribution was the systematic use of experiments to prove his theories, which is the core of the modern scientific method.

Answer: B

Explanation: He disproved the "Emission Theory" held by the Greeks (Pythagoras and Euclid), which suggested eyes emitted rays like a flashlight.

Answer: C

Explanation: He used the term Camera Obscura (meaning "dark room") to describe the phenomenon of light passing through a pinhole to form an image.

Part 2: True or False (Answer Key)

False

Explanation: Ibn al-Haytham was the first to argue that vision is a process that is completed in the brain, not just the eye.

True

Explanation: Much of his groundbreaking work was done during his ten years of house arrest after the failed Nile project.

False

Explanation: His work was translated into Latin and was highly influential to Renaissance figures like Leonardo da Vinci and Johannes Kepler.

Part 3: Critical Thinking (Guided Answers)

The "Enemy" of the Text: Students should discuss the importance of fact-checking and skepticism. Just as Ibn al-Haytham didn't believe ancient texts without proof, we shouldn't believe a headline without looking for primary sources, evidence, or seeing if the "experiment" (the event) can be verified by others.

Failure as a Catalyst: This answer should focus on resilience. Instead of giving up when his engineering project failed or when he was imprisoned, he used the forced silence and darkness of his room to observe light. He turned a negative situation into a "laboratory" for his greatest work.

The Anatomy of Vision: Light travels in straight lines. When light from the top of an object passes through a small opening (like the pupil or a pinhole), it hits the bottom of the screen (or retina). Conversely, light from the bottom hits the top.

The Brain's Role: Since the image on our retina is inverted, our brain must interpret and "flip" the signal so we perceive the world right-side up.

Ibn al-Haytham was the First Scientist

"I constantly sought knowledge and truth, and it became my belief that for gaining access to the effulgence and closeness to God, there is no better way than that of searching for truth and knowledge." Ibn al-Haytham

Before Ibn al-Haytham, scholars relied mainly on logical reasoning. He realized that logic alone was insufficient and insisted that ideas must be tested through physical experiments—what He called "true demonstrations." This approach laid the foundation for the scientific method, which is now central to all scientific inquiry.

Ibn al-Haytham applied this rigorous methodology to his own work, discarding earlier treatises that lacked experimental proof. His insistence on evidence and experimentation inspired generations of scientists. This biography highlights both his discoveries and the personal journey that led him to become the first modern scientist.

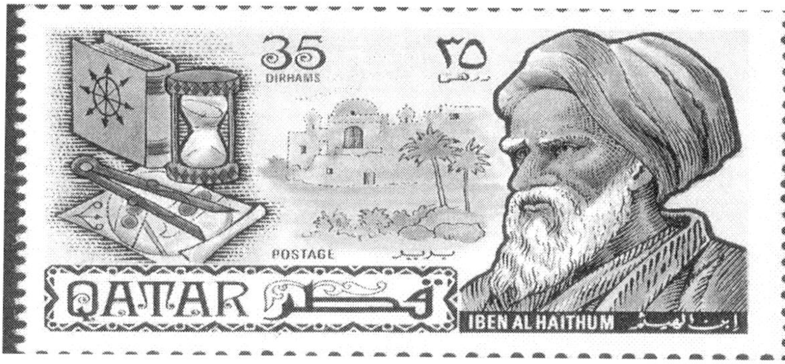

Postage stamp issued by the State of Qatar (1971) representing Ibn al-Haytham and his scientific achievements celebrating the 1,000th anniversary of Ibn al-Haytham (Alhazen). Reproduced for educational purposes.

Before Ibn al-Haytham, scholars relied mainly on logical reasoning. He realized that logic alone was insufficient and insisted that ideas must be tested through physical experiments—what he called "true demonstrations." This approach laid the foundation for the scientific method, which is now central to all scientific inquiry.

Ibn al-Haytham applied this rigorous methodology to his own work, discarding earlier treatises that lacked experimental proof. His insistence on evidence and experimentation inspired generations of scientists. This biography highlights both his discoveries and the personal journey that led him to become the first modern scientist.

Ibn Hayutham stated that Truth is sought for its own sake ... Finding the truth is difficult, and the road to it is rough. For the truth, they are plunged into obscurity. God, however, has not preserved the scientist from error and has not safeguarded science from shortcomings and faults. We need to test and retest scientists' findings to find the truth.

"The seeker after the truth is not one who studies the writings of the ancients... but rather the one who suspects his faith in them and questions what he gathers from them."